ANIMAL CAMOUFLAGE CLASH

RACHAEL L. THOMAS

Consulting Editor, Diane Craig, M.A./Reading Specialist

Super Sandcastle

An Imprint of Abdo Publishing
abdobooks.com

abdobooks.com

Published by Abdo Publishing, a division of ABDO, PO Box 398166, Minneapolis, Minnesota 55439. Copyright © 2020 by Abdo Consulting Group, Inc. International copyrights reserved in all countries. No part of this book may be reproduced in any form without written permission from the publisher. Super Sandcastle™ is a trademark and logo of Abdo Publishing.

Printed in the United States of America, North Mankato, Minnesota
102019
012020

Design: Sarah DeYoung, Mighty Media, Inc.
Production: Mighty Media, Inc.
Editor: Jessica Rusick
Cover Photographs: Shutterstock Images
Interior Photographs: Getty Images/iStockphoto, pp. 5 (bottom left), 23; Shutterstock Images, pp. 4, 5, 6, 7, 8, 9, 10, 11, 12, 13, 14, 15, 16, 17, 18, 19, 20, 21, 22

Library of Congress Control Number: 2019943208

Publisher's Cataloging-in-Publication Data
Names: Thomas, Rachael L., author.
Title: Animal camouflage clash / by Rachael L. Thomas
Description: Minneapolis, Minnesota : Abdo Publishing, 2020 | Series: Incredible animal face-offs
Identifiers: ISBN 9781532191930 (lib. bdg.) | ISBN 9781532178733 (ebook)
Subjects: LCSH: Animal camouflage--Juvenile literature. | Animal defense mechanisms--Juvenile literature.
 | Animals, Habits and behavior of--Juvenile literature. | Social behavior in animals--Juvenile literature.
Classification: DDC 591.572--dc23

Super Sandcastle™ books are created by a team of professional educators, reading specialists, and content developers around five essential components—phonemic awareness, phonics, vocabulary, text comprehension, and fluency—to assist young readers as they develop reading skills and strategies and increase their general knowledge. All books are written, reviewed, and leveled for guided reading, early reading intervention, and Accelerated Reader™ programs for use in shared, guided, and independent reading and writing activities to support a balanced approach to literacy instruction.

CONTENTS

Battle of the Camouflage Artists 4

Disguiser Duel 6

Mimicry Match-Up 12

Show-Off Showdown 18

Glossary 24

BATTLE OF THE CAMOUFLAGE ARTISTS

LEAFY SEA DRAGON

ZEBRA

The animal kingdom is full of stars. But some animals stand out. These animals have the best **camouflage**.

Animals use camouflage to blend in with their surroundings. This helps them hide from predators and prey.

Camouflage champions are all around us. But what if you matched them up in face-offs? Which animal would have the best camouflage?

DISGUISER DUEL

Arctic foxes look like their surroundings. Leafy sea dragons do too! But which animal would win a disguiser duel?

Thick layer of body fat for warmth

ARCTIC FOX
WINTRY WHITE
This animal's fur helps it blend in with snow and ice. Say hello to the Arctic fox!

Furry soles to grip ice

White fur to match surroundings

ARCTIC FOX STATS

HOME
The entire Arctic **tundra**, including Alaska, Canada, and Iceland

FOOD
Rodents, birds, and fish

SIZE
Up to 3½ feet (1 m) long, including a 12-inch (30 cm) tail

AN ARCTIC FOX IS LONGER THAN A MICROWAVE.

ARCTIC FOX

MICROWAVE

Small fins for swimming

Wavy, leaflike body parts to blend in with seaweed

LEAFY SEA DRAGON

LEAFY LOOK-A-LIKE

This sea creature grows extra body parts to match its underwater home. Make some noise for the leafy sea dragon!

SEA DRAGON STATS

HOME
Waters off the southern and eastern coasts of Australia

FOOD
Plankton, shrimp, and other small **crustaceans**

SIZE
Up to 17 inches (43 cm) long

A LEAFY SEA DRAGON IS ABOUT AS LONG AS AN OPEN BOOK.

LEAFY SEA DRAGON   BOOK

DISGUISER DUEL
ARCTIC FOX VS LEAFY SEA DRAGON

ARCTIC FOX

An Artic fox's fur is white in winter. It matches the Arctic snow and ice!

THAT'S SNEAKY!
The Arctic fox's **camouflage** helps it sneak up on prey. The foxes hunt birds and **rodents**. They also hunt fish.

CLIMATE CAMOUFLAGE
An Arctic fox's camouflage works best in snow. The less snow there is, the fewer places Arctic foxes can safely live.

LEAFY SEA DRAGON

Leafy sea dragons look like seaweed. They move like it too! A leafy sea dragon's body sways in ocean currents.

THAT'S SNEAKY!

Leafy sea dragons are **fragile**. So, they need **camouflage** for protection.

CLIMATE CAMOUFLAGE

Human pollution can harm seaweed. Without seaweed, sea dragons can't survive.

MIMICRY MATCH-UP

The Indonesian **mimic** octopus uses **camouflage** to adapt to different situations. The decorator crab does too! But which animal would win in a mimicry match-up?

Boneless legs and body to squeeze into small places

Brown and white striped skin can change color

MIMIC OCTOPUS
CONFIDENT COSTUMER

This eight-legged animal can copy other sea creatures. Give it up for the **mimic** octopus!

Suckers on arms to stick to ocean rocks and grab prey

MIMIC OCTOPUS STATS

HOME
Mostly spotted off the coast of Indonesia

FOOD
Worms, crabs, small fish, and other mimic octopuses

SIZE
Up to 24 inches (61 cm) long

A MIMIC OCTOPUS IS AS LONG AS AN UMBRELLA.

MIMIC OCTOPUS — UMBRELLA

Small hooks to which sea objects can stick

DECORATOR CRAB
SHELLED STYLIST

This **crustacean** uses underwater items to become part of its surroundings. Get on your feet for the decorator crab!

Hard shell for protection

DECORATOR CRAB STATS

HOME
All over the world

FOOD
Small crustaceans, sponges, and **algae**

SIZE
Up to 5 inches (13 cm) long

A DECORATOR CRAB IS AS WIDE AS A DESSERT PLATE.

DECORATOR CRAB

DESSERT PLATE

MIMICRY MATCH-UP
MIMIC OCTOPUS VS DECORATOR CRAB

MIMIC OCTOPUS

Mimic octopuses copy other sea animals. The octopuses can copy an animal's colors and shape.

THAT'S VERSATILE!
A mimic octopus can copy a sea snake. It does this by digging into the ocean floor and waving its arms. Sea snakes are deadly. So, predators stay away.

KILLER CAMOUFLAGE
Mimic octopuses also mimic their prey. Once prey gets close, the octopus strikes!

DECORATOR CRAB

Decorator crabs attach objects to their shells. This helps the crabs blend in with their surroundings.

THAT'S VERSATILE!

There are many different types of decorator crabs. Seaweed crabs attach **algae** to their shells. Sponge crabs attach sea sponges.

KILLER CAMOUFLAGE

Decorator crabs may use harmful objects for **camouflage**. This does not hurt the crab. It protects the crab from predators!

SHOW-OFF SHOWDOWN

Zebras use **camouflage** to confuse other animals. Scarlet kingsnakes do too! But which animal would win in a show-off showdown?

ZEBRA
STRIPED SHOW-OFF
This African animal can ward off pests with its stripes. Say hello to the zebra!

Bold, black-and-white stripes to confuse pests

Long, thin tail to swat flies

ZEBRA STATS

HOME
Grasslands and woodlands in eastern and southern Africa

FOOD
Grass, leaves, and twigs

SIZE
Up to 5 feet (1.5 m) tall at the shoulder

A ZEBRA IS SHORTER THAN AN AVERAGE MAN.

ZEBRA   MAN

SCARLET KINGSNAKE
CLEVER COPYCAT

This snake looks and moves like some deadly snakes. But it is actually harmless! Make some noise for the scarlet kingsnake!

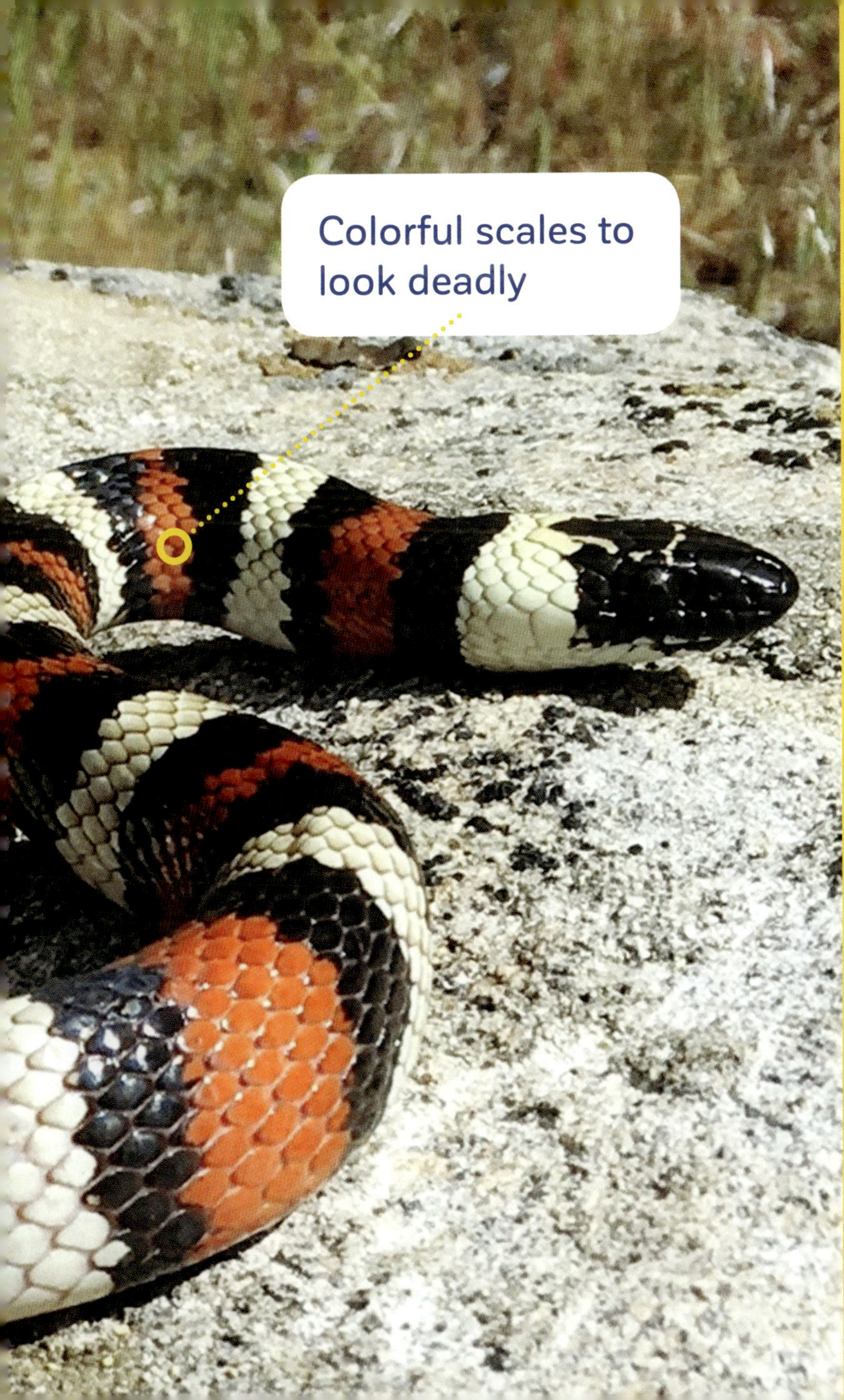

Colorful scales to look deadly

SCARLET KINGSNAKE STATS

HOME
Virginia to Florida, and as far west as the Mississippi River

FOOD
Frogs, toads, lizards, and other small snakes

SIZE
About 12 to 24 inches (30 to 61 cm) long

A SCARLET KINGSNAKE IS LONGER THAN A LOAF OF BREAD.

SCARLET KINGSNAKE   LOAF OF BREAD

21

SHOW-OFF SHOWDOWN
ZEBRA VS SCARLET KINGSNAKE

ZEBRA

A zebra's stripes keep flies away. Stripes make a zebra less likely to get bitten.

MASTERS OF DISGUISE
Scientists believe zebra stripes create an **illusion**. The stripes look like they are moving. This makes it hard for flies to land.

THAT'S COOL!
Each zebra's stripes are different. They are like human fingerprints!

SCARLET KINGSNAKE

A scarlet kingsnake is colored like a coral snake. Coral snakes are deadly. So, predators stay away.

MASTERS OF DISGUISE
Scarlet kingsnakes also shake their tails. So, they look like deadly rattlesnakes.

THAT'S COOL!
Scarlet kingsnakes look harmful. But they are not. They are even popular pets!

GLOSSARY

algae—a plant or plantlike organism that lives in water.

camouflage—the features of an animal that make it look like its surroundings.

crustacean—a sea creature, such as a lobster, crab, or shrimp, that has a hard, external skeleton.

fragile—easily broken or damaged.

illusion—something that looks real but is not.

mimic—to copy.

plankton—very tiny organisms that live in water.

rodent—a mammal with large, sharp front teeth, such as a rat, mouse, or squirrel.

tundra—a large area of flat land in northern parts of the world where there are no trees and the ground is always frozen.